Shapes and Solids

by Michelle Schaub

Table of Contents

What Are Shapes?................page 4

What Are Solids?................page 10

What Solids Do You Know?........page 12

I need to know these words.

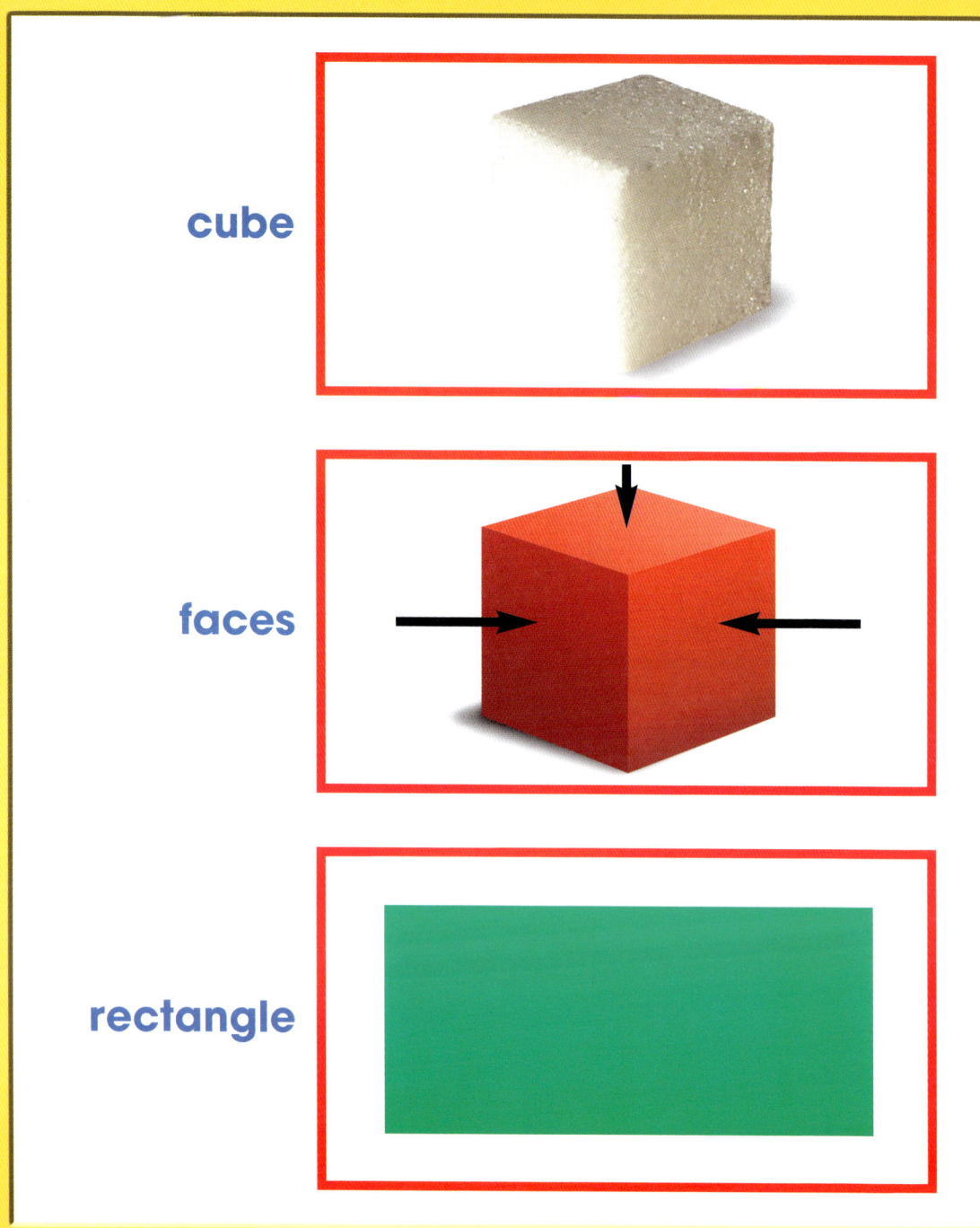

cube

faces

rectangle

shapes

solids

square

What Are Shapes?

Look at the checkerboard. Some parts are red. Some parts are black.
All of the parts are shapes.
The shapes are squares.

▲ The checkerboard is made of squares.

All shapes are flat. You can draw a shape. You can draw a square.

▲ You can trace a square.

A square is a shape. A square has four sides. The sides are the same length.

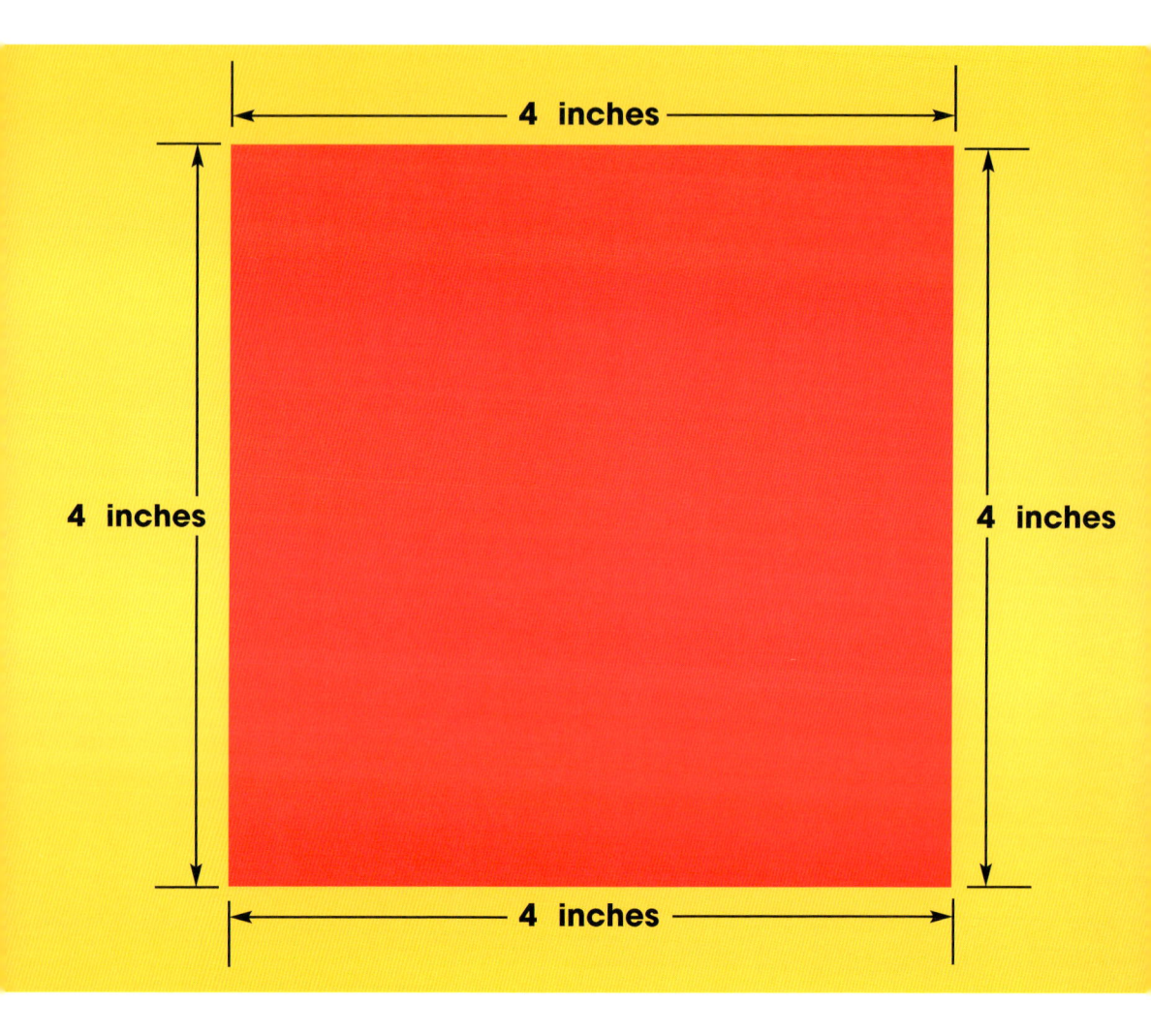

4 inches

4 inches

4 inches

4 inches

▲ The sides of a square have equal lengths.

A square has corners.
A square has four corners.

▲ This square has four corners.

A rectangle is another shape.
A rectangle has four sides.
A rectangle also has four corners.

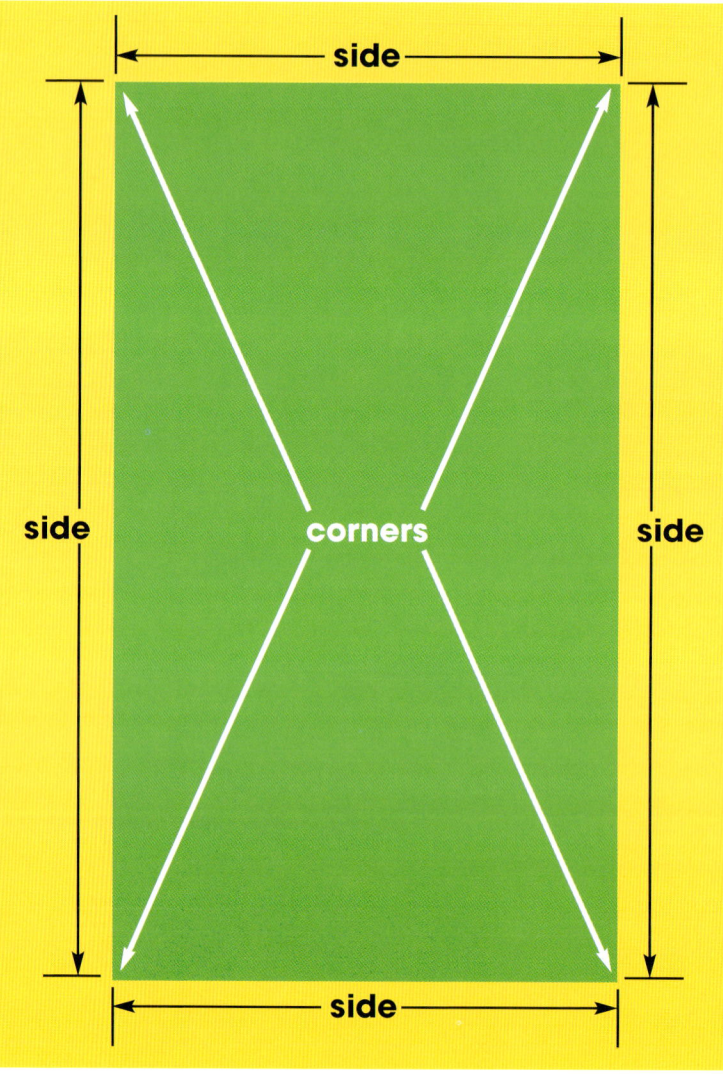

▲ A rectangle has four sides.

A rectangle has two long sides.
The long sides are the same length.
A rectangle has two short sides.
The short sides are the same length.

▲ The face of this door is a rectangle.

What Are Solids?

Look at the box. The box is a cube. A cube is a solid. You can pick up a solid.

▲ This is a cube.

The sides of the cube are called faces. You can trace the faces of the cube. The faces are squares. The squares are flat.

face

▲ A cube has six faces.

What Solids Do You Know?

Look at the sugar cube. A sugar cube is a solid. A sugar cube has flat parts. The flat parts are faces.

▲ How many faces can you see?

A cube has six faces. All the faces are the same size. All the faces are the same shape. Each face is a square.

▲ You can count the faces on the cubes.

This girl holds a box. The box is another solid.

▲ The box is a solid.

The box has six faces. Each face is a rectangle. How is the box different from a cube?

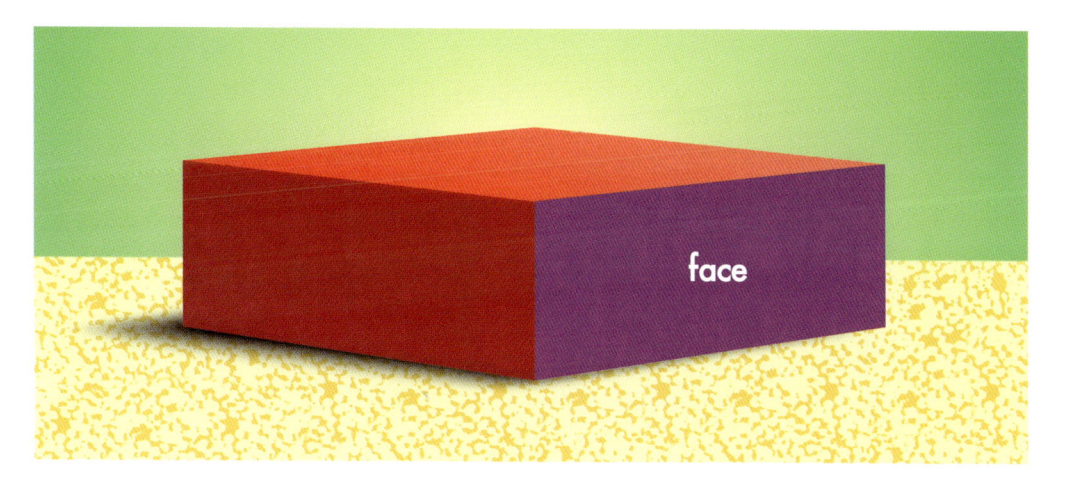

face

face

▲ The faces of this box are not squares.

Look at the pictures. How many shapes do you see? How many solids do you see?

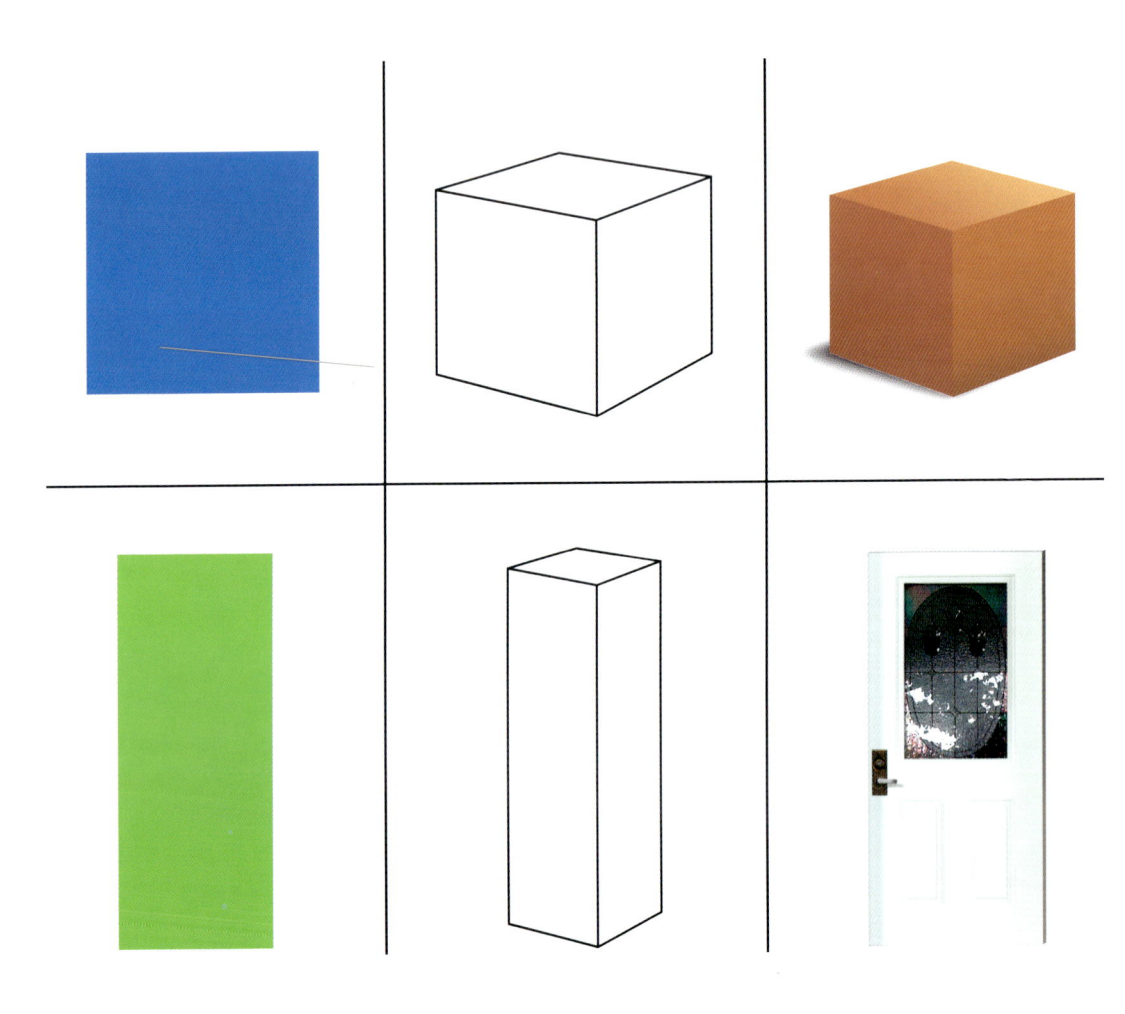